Table of Contents

1.0 Introduction 1

2.0 Petroleum Refining 1

    2.1   **Overview of Petroleum Refining Industry** 1

    2.2   **Petroleum Refining GHG Emission Sources** 3

        2.2.1   Stationary Combustion Sources 5

        2.2.2   Flares 6

        2.2.3   Catalytic Cracking Units 6

        2.2.4   Coking Units 7

        2.2.5   Catalytic Reforming Units 8

        2.2.6   Sulfur Recovery Vents 9

        2.2.7   Hydrogen Plants 9

        2.2.8   Asphalt Blowing Stills 9

        2.2.9   Storage Tanks 10

        2.2.10 Coke Calcining Units 10

        2.2.11 Other Ancillary Sources 10

3.0 Summary of GHG Reduction Measures 11

4.0 Energy Programs and Management Systems 16

    4.1   **Sector-Specific Plant Performance Benchmarks** 19

    4.2   **Industry Energy Efficiency Initiatives** 19

    4.3   **Energy Efficiency Improvements in Facility Operations** 19

        4.3.1   Monitoring and Process Control Systems 19

        4.3.2   High Efficiency Motors 20

        4.3.3   Variable Speed Drives 21

        4.3.4   Optimization of Compressed Air Systems 21

        4.3.5   Lighting System Efficiency Improvements 21

5.0 GHG Reduction Measures by Source 22

    5.1   **Stationary Combustion Sources** 22

        5.1.1   Steam Generating Boilers 22

        5.1.2   Process Heaters 24

        5.1.3   Combined Heat and Power (CHP) 24

        5.1.4   Carbon Capture 25

    5.2   **Fuel Gas Systems and Flares** 26

        5.2.1   Fuel Gas Systems 26

        5.2.2   Flares 27

    5.3   **Cracking Units** 28

        5.3.1   Catalytic Cracking Units 28

        5.3.2   Hydrocracking Units 29

    5.4   **Coking Units** 29

        5.4.1   Fluid Coking Units 29

        5.4.2   Flexicoking Units 30

        5.4.3   Delayed Coking Units 30

    5.5   **Catalytic Reforming Units** 31

    5.6   **Sulfur Recovery Units** 31

5.7   **Hydrogen Production Units** **31**
5.7.1   Combustion Air and Feed/Steam Preheat 32
5.7.2   Cogeneration 32
5.7.3   Hydrogen Purification 32
5.8   Hydrotreating Units **32**
5.9   Crude Desalting and Distillation Units **33**
5.9.1   Desalter Design 33
5.9.2   Progressive Distillation Design 33
5.10   Storage Tanks **34**
5.10.1 Vapor Recovery or Control for Unstabilized Crude Oil Tanks 34
5.10.2 Heated Storage Tank Insulation 35

6.0 References **35**
        EPA Contact 38

List of Figures

Figure 1. Simplified flowchart of refining processes and product flows. Adapted from Gary and Handwerk (1994). 2
Figure 2. Contribution of different emission sources to the nationwide CO_2 equivalent GHG emissions from petroleum refineries. 3
Figure 3. Contribution of different GHG to the nationwide CO_2 equivalent GHG emissions from petroleum refineries. 4
Figure 4. Direct CO_2 emissions from fuel consumption and indirect CO_2 emissions electricity and steam purchases at U.S. petroleum refineries from 2003 to 2008 5
Figure 5. ENERGY STAR Guidelines for Energy Management 18
Figure 6. Process schematic of a progressive distillation process (from ARCADIS, 2008). 34

List of Tables

Table 1. Summary of GHG Reduction Measures for the Petroleum Refining Industry 11

1.0 Introduction

This document is one of several white papers that summarize readily available information on control techniques and measures to mitigate greenhouse gas (GHG) emissions from specific industrial sectors. These white papers are solely intended to provide basic information on GHG control technologies and reduction measures in order to assist States and local air pollution control agencies, tribal authorities, and regulated entities in implementing technologies or measures to reduce GHGs under the Clean Air Act, particularly in permitting under the prevention of significant deterioration (PSD) program and the assessment of best available control technology (BACT). These white papers do not set policy, standards or otherwise establish any binding requirements; such requirements are contained in the applicable EPA regulations and approved state implementation plans.

This document provides information on control techniques and measures that are available to mitigate greenhouse gas (GHG) emissions from the petroleum refining industry at this time. Because the primary GHG emitted by the petroleum refining industry are carbon dioxide (CO_2) and methane (CH_4), the control technologies and measures presented here focus on these pollutants. While a large number of available technologies are discussed here, this paper does not necessarily represent all potentially available technologies or measures that that may be considered for any given source for the purposes of reducing its GHG emissions. For example, controls that are applied to other industrial source categories with exhaust streams similar to the petroleum refining industry may be available through "technology transfer" or new technologies may be developed for use in this sector.

The information presented in this document does not represent U.S. EPA endorsement of any particular control strategy. As such, it should not be construed as EPA approval of a particular control technology or measure, or of the emissions reductions that could be achieved by a particular unit or source under review.

2.0 Petroleum Refining

2.1 Overview of Petroleum Refining Industry

Petroleum refineries produce liquefied petroleum gases (LPG), motor gasoline, jet fuels, kerosene, distillate fuel oils, residual fuel oils, lubricants, asphalt (bitumen), and other products through distillation of crude oil or through redistillation, cracking, or reforming of unfinished petroleum derivatives. There are three basic types of refineries: topping refineries, hydroskimming refineries, and upgrading refineries (also referred to as "conversion" or "complex" refineries). Topping refineries have a crude distillation column and produce naphtha and other intermediate products, but not gasoline. There are only a few topping refineries in the U.S., predominately in Alaska. Hydroskimming refineries have mild conversion units such as hydrotreating units and/or reforming units to produce finished gasoline products, but they do not upgrade heavier components of the crude oil that exit near the bottom of the crude distillation column. Some topping/hydroskimming refineries specialize in processing heavy crude oils to produce asphalt. There are eight operating asphalt plants and approximately 20 other hydroskimming refineries operating in the United States as of January 2006 (Energy Information Administration [EIA], 2006). The vast majority (approximately 75 to 80 percent) of the approximately 150 domestic refineries are upgrading/conversion refineries. Upgrading/conversion refineries have cracking or coking operations to convert long-chain, high molecular weight hydrocarbons ("heavy distillates") into smaller hydrocarbons that can be used to produce gasoline product ("light distillates") and other higher value products and petrochemical feedstocks.

Figure 1 provides a simplified flow diagram of a typical refinery. The flow of intermediates between the processes will vary by refinery, and depends on the structure of the refinery, type of crude processes, as well as product mix. The first process unit in nearly all refineries is the crude oil or "atmospheric" distillation unit (CDU). Different conversion processes are available using thermal or catalytic processes, *e.g.*, delayed coking, catalytic cracking, or catalytic reforming, to produce the desired mix of products from the crude oil. The products may be treated to upgrade the product quality (*e.g.*, sulfur removal using a hydrotreater). Side processes that are used to condition inputs or produce hydrogen or by-products include crude conditioning (*e.g.*, desalting), hydrogen production, power and steam production, and asphalt production. Lubricants and other specialized products may be produced at special locations. More detailed descriptions of petroleum refining processes are available in other locations (U.S. EPA, 1995, 1998; U.S. Department of Energy [DOE], 2007).

2.2   *Petroleum Refining GHG Emission Sources*

The petroleum refining industry is the nation's second-highest industrial consumer of energy (U.S. DOE, 2007). Nearly all of the energy consumed is fossil fuel for combustion; therefore, the petroleum refining industry is a significant source of GHG emissions. In addition to the combustion-related sources (*e.g.*, process heaters and boilers), there are certain processes, such as fluid catalytic cracking units (FCCU), hydrogen production units, and sulfur recovery plants, which have significant process emissions of CO_2. Methane emissions from a typical petroleum refinery arise from process equipment leaks, crude oil storage tanks, asphalt blowing, delayed coking units, and blow down systems. Asphalt blowing and flaring of waste gas also contributes to the overall CO_2 and CH_4 emissions at the refinery. Based on a bottom-up, refinery-specific analysis (adapted from Coburn, 2007, and U.S. EPA, 2008), GHG emissions from petroleum refineries were estimated to be 214-million metric tons of CO_2 equivalents (CO_2e), based on production rates in 2005. **Figure 2** provides a breakdown of the nationwide emissions projected for different parts of the petroleum refineries based on this bottom-up analysis.

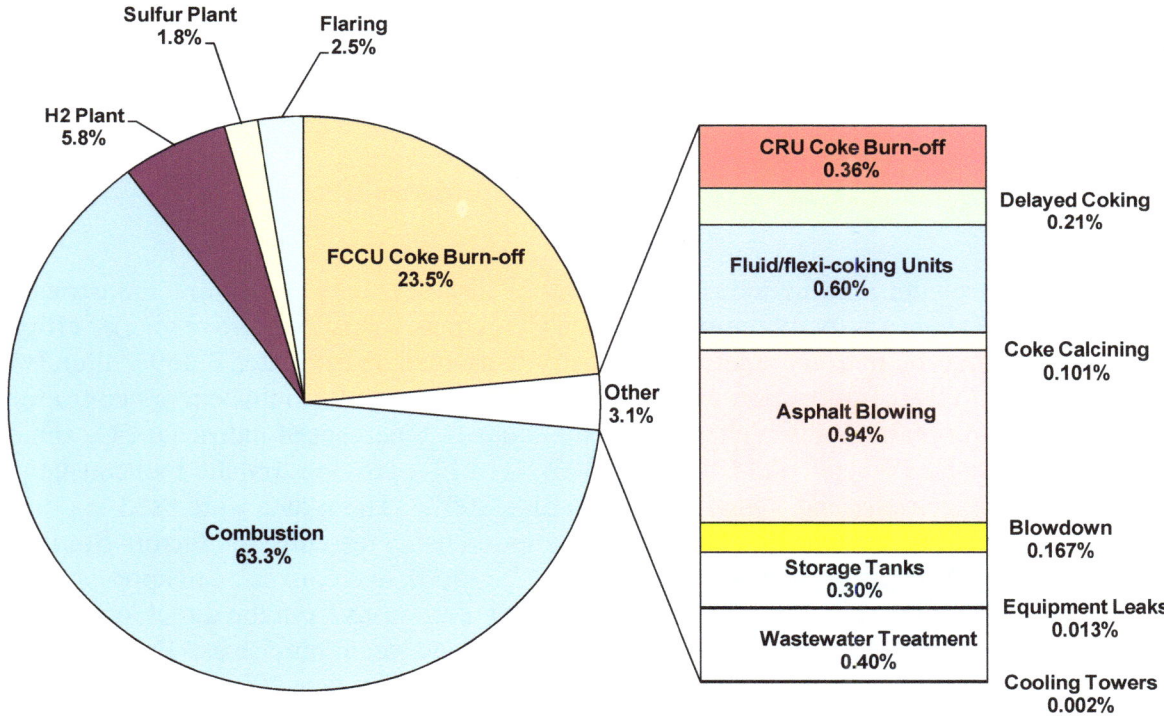

Figure 2. Contribution of different emission sources to the nationwide CO_2 equivalent GHG emissions from petroleum refineries.

Figure 3 presents what GHG are emitted by refineries. CO_2 is the predominant GHG emitted by petroleum refineries, accounting for almost 98 percent of all GHG emissions at petroleum refineries. Methane emissions are 4.7-million metric tons CO_2e and account for 2.25 percent of the petroleum refinery emissions nationwide. Note that the relative magnitude of CO_2 and CH_4 emissions is dependent on the types of process units and other characteristics of the refinery. Facilities that do not have catalytic cracking units and hydrogen plants will tend to have a higher fraction of their total GHG emissions released as CH_4.

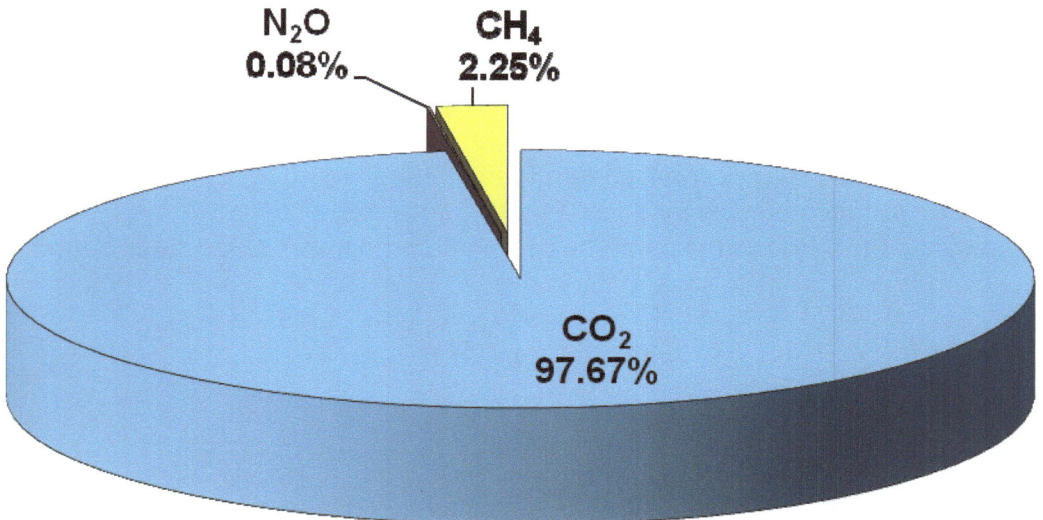

Figure 3. GHG emissions from petroleum refineries.

The petroleum refining industry is one of the largest energy consuming industries in the United States. Thus, a primary option to reduce CO_2 emissions is to improve energy efficiency. In 2001, the domestic petroleum refining industry consumed an estimated 3,369 trillion British Thermal Units (TBtu). One report estimated the CO_2 emissions from this energy consumption to be about 222 million tonnes, which accounts for about 11.6 percent of industrial CO_2 emissions in the United States (Worrell and Galitsky, 2005). The EIA provides on-site fuel consumption data as well as electricity and steam purchases (EIA, 2009). These data were used to estimate the CO_2 emissions resulting from this fuel consumption using the emission factors from the Intergovernmental Panel on Climate Change (IPCC) (2006), and converted to appropriate units (Coburn, 2007). **Figure 4** presents the projected CO_2 emissions from the direct, on-site fuel consumption, as well as the indirect, off-site electricity and steam purchases. From Coburn (2007), the on-site annual CO_2 emissions from fuel combustion were 190 million tonnes in 2005 and the overall CO_2 emissions from energy consumption (including purchased steam and electricity) were 216 million tonnes in 2005, which agrees well with the estimate of Worrell (2005). As seen in Figure 4, catalyst coke consumption dropped 10 percent from 2006 to 2008. Much of the resulting CO_2 emission reductions were offset by increased electricity and steam purchases. As nearly all catalytic cracking units recover the latent heat from the catalyst coke burn-off exhaust to produce steam and/or electricity, the decrease in catalyst coke consumption does not translate into an equivalent net GHG emissions reduction when indirect CO_2 emissions from electricity and steam purchases are considered.

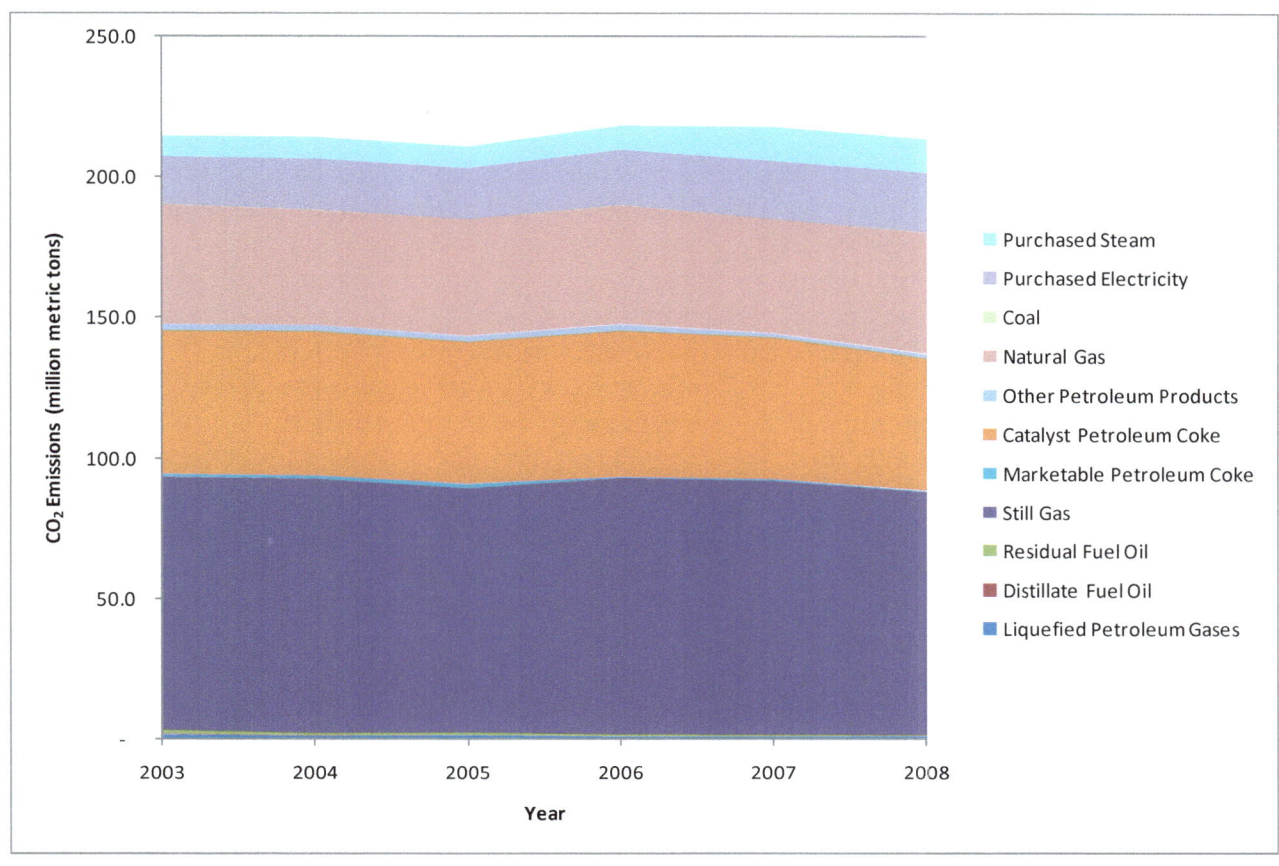

Figure 4. Direct CO$_2$ emissions from fuel consumption and indirect CO$_2$ emissions from electricity and steam purchases at U.S. petroleum refineries from 2003 to 2008.

The remainder of this section provides brief descriptions of the process units and other sources that generate significant GHG emissions at a petroleum refinery.

2.2.1 Stationary Combustion Sources

Stationary combustion sources are the largest sources of GHG emissions at a petroleum refinery. Combustion sources primarily emit CO$_2$, but they also emit small amounts of CH$_4$ and nitrous oxide (N$_2$O). Stationary combustion sources at a petroleum refinery include process heaters, boilers, combustion turbines, and similar devices. For this document, flares are considered a distinct emission source separate from other stationary combustion sources. Nearly all refinery process units use process heaters. Typically, the largest process heaters at a petroleum refinery are associated with the crude oil atmospheric and vacuum distillation units and the catalytic reforming unit (if present at the refinery).

In addition to direct process heat, many refinery processes also have steam and electricity requirements. Some refineries purchase steam to meet their process's steam requirements; others use dedicated on-site boilers to meet their steam needs. Similarly, some refineries purchase electricity from the grid to run their pumps and other electrical equipment; other refineries have co-generation facilities to meet their electricity needs and may produce excess electricity to sell to the grid. Refineries that produce their own steam or electricity will have higher on-site fuel usage, all other factors being equal, than refineries that purchase these utilities. A boiler for producing plant steam can be the largest source of GHG emissions at the refinery, particularly at refineries that do not have catalytic cracking units.

The predominant fuel used at petroleum refineries is refinery fuel gas (RFG), which is also known as still gas. RFG is a mixture of light C1 to C4 hydrocarbons, hydrogen, hydrogen sulfide (H_2S), and other gases that exit the top (overhead) of the distillation column and remain uncondensed as they pass through the overhead condenser. RFG produced at different locations within the refinery is typically compressed, treated to remove H_2S (if necessary), and routed to a central location (*i.e.*, mix drum) to supply fuel to the various process heaters at the refinery. This RFG collection and distribution system is referred to as the fuel gas system. A refinery may have several fuel gas systems, depending on the configuration of the refinery, supplying fuel to different process heaters and boilers.

The fuel gas generated at the refinery is typically augmented with natural gas to supply the full energy needs of the refinery. Depending on the types of crude oil processed and the process units in operation, the amount of supplemental natural gas needed can change significantly. Consequently, there may be significant variability in the fuel gas composition between different refineries and even within a refinery as certain units are taken off-line for maintenance.

2.2.2   Flares

Flares are commonly used in refineries as safety devices to receive gases during periods of process upsets, equipment malfunctions, and unit start-up and shutdowns. Some flares receive only low flows of "purge" or "sweep" gas to prevent air (oxygen) from entering the flare header and possibly the fuel gas system while maintaining the readiness of the flare in the event of a significant malfunction or process upset. Some flares may receive excess process gas on a frequent or routine basis. Some flares may be used solely as control devices for regulatory purposes. Combustion of gas in a flare results in emissions of predominately CO_2, along with small amounts of CH_4 and N_2O.

2.2.3   Catalytic Cracking Units

In the catalytic cracking process, heat and pressure are used with a catalyst to break large hydrocarbons into smaller molecules. The FCCU is the most common type of catalytic cracking unit currently in use. The FCCU feed is pre-heated to between 500 and 800 degrees Fahrenheit (°F) and contacted with fine catalyst particles from the regenerator section, which are at about 1,300 °F in the feed line ("riser"). The feed vapor, which is heavy distillate oil from the crude or vacuum distillation column, reacts when contacted with the hot catalyst to break (or crack) the large hydrocarbon compounds into a variety of lighter hydrocarbons. During this cracking process, coke is deposited on the catalyst particles, which deactivates the catalyst. The catalyst separates from the reacted ("cracked") vapors in the reactor; the vapors continue to a fractionation tower and the catalyst is recycled to the regenerator portion of the FCCU to burn-off the coke deposits and prepare the catalyst for reuse in the FCCU riser/reactor (U.S. EPA, 1998).

The FCCU catalyst regenerator generates GHG through the combustion of coke (essentially solid carbon with small amounts of hydrogen and various impurities) that was deposited on the catalyst particles during the cracking process. CO_2 is the primary GHG emitted; small quantities of CH_4 and N_2O are also emitted during "coke burn-off." An FCCU catalyst regenerator can be designed for complete or partial combustion. A complete-combustion FCCU operates with sufficient air to convert most of the carbon to CO_2 rather than carbon monoxide (CO). A partial-combustion FCCU generates CO as well as CO_2, so most partial-combustion FCCU are typically followed by a CO boiler to convert the CO to CO_2. Most refineries that operate an FCCU recover useful heat generated from the combustion of catalyst coke during catalyst regeneration; the heat recovered from catalyst coke combustion offsets some of the refinery's ancillary energy needs. The FCCU catalyst regeneration or coke burn-off vent is often the largest single source of GHG emissions at a refinery.

Thermal catalytic cracking units (TCCU) are similar to FCCU, except that the catalyst particles are much bigger and the system uses a moving bed reactor rather than a fluidized system. The generation of GHG, however, is the same. Specifically, GHG are generated in the regenerator section of the TCCU when coke deposited on the catalyst particles is burned-off in order to restore catalyst activity.

2.2.4 Coking Units

Coking is another cracking process, usually used at a refinery to generate transportation fuels, such as gasoline and diesel, from lower-value fuel oils. A desired by-product of the coking reaction is petroleum coke, which can be used as a fuel for power plants as well as a raw material for carbon and graphite products. Coking units are often installed at existing refineries to increase the refinery's ability to process heavier crude oils. There are three basic types of coking units: delayed coking units, (traditional) fluid coking units, and flexicoking units. Delayed coking units are the most common, and all new coking units are expected to be delayed cokers.

Delayed Coking Units. Delayed coking is a semibatch process using two coke drums and a single fractionator tower (distillation column) and coking furnace. A feed stream of heavy residues is introduced to the fractionating tower. The bottoms from the fractionator are heated to about 900 to 1,000 °F in the coking furnace, and then fed to an insulated coke drum where thermal cracking produces lighter (cracked) reaction products and coke. The reaction products produced in the coke drum are fed back to the fractionator for product separation. After the coke drum becomes filled with coke, the feed is alternated to the parallel (empty) coke drum, and the filled coke drum is purged and cooled, first by steam injection, and then by water addition. A coke drum blowdown system recovers hydrocarbon and steam vapors generated during the quenching and steaming process. Once cooled, the coke drum is vented to the atmosphere, opened, and then high pressure water jets are used to cut the coke from the drum. After the coke cutting cycle, the drum is closed and preheated to prepare the vessel for going back on-line (i.e., receiving heated feed). A typical coking cycle will last for 16 to 24 hours on-line and 16 to 24 hours cooling and decoking. The primary GHG released from a delayed coking unit is CH_4, which is emitted both from the blowdown system (if not controlled) and from the atmospheric venting and opening of the coke drum.

Fluid Coking Units. The fluid coking process is continuous and occurs in a reactor rather than a coke drum like the delayed coking process. Fluid coking units produce a higher grade of petroleum coke than delayed coking units; however, unlike delayed coking units that use large process preheaters, fluid coking units burn 15 to 25 percent of the coke produced to provide the heat needed for the coking reactions (U.S. DOE, 2007). The coke is burned with limited air, so large quantities of CO are produced (similar to a partial combustion FCCU), which are subsequently burned in a CO boiler. Like the FCCU, the combustion of the petroleum coke and subsequent combustion of CO generates large quantities of CO_2 along with small amounts of CH_4 and N_2O. For the few refineries with fluid coking units, the fluid coking units are significant contributors to the refinery's GHG emissions. Fluid coking units are not significant contributors to the nationwide emissions totals because there are only three fluid coking units in the United States; however, fluid coking units have emissions comparable to (and slightly greater than) catalytic cracking units of the same throughput capacity.

Flexicoking Units. The flexicoking process is very similar to the fluid coking unit except that a coke gasifier is added that burns nearly all of the produced coke at $1700 - 1800\ °F$ with steam to produce low heating value synthesis gas (syngas). The produced syngas, along with entrained fines, is routed through the heater vessel for fluidization of the hot coke bed and for heat transfer to the solids. The syngas is then treated to remove entrained particles and reduced sulfur compounds and the syngas can then be used in specially designed boilers or other combustion sources that can accommodate the low heat content of the syngas. Most of the CO_2 emissions produced in the flexicoking unit will not be released at the unit, but rather it will be part of the syngas. Some of the CO_2 produced in the flexicoking unit is expected to be removed as part of the sulfur removal process and subsequently released in the sulfur recovery plant; the CO_2 that remains in the scrubbed syngas will be released from the stationary combustion unit that uses the syngas as fuel (usually a boiler specifically designed to use the low heating value content syngas). Therefore, while the flexicoking unit is not expected to have significant GHG emissions directly from the unit, the flexicoking unit will impact the energy balance and GHG emissions from other sources at the refinery.

2.2.5 Catalytic Reforming Units

In the catalytic reforming unit (CRU), low-octane hydrocarbon distillates, generally gasoline and naphtha are reacted with a catalyst to produce aromatic compounds such as benzene. An important by-product of the reforming reaction is hydrogen. The feed to the CRU must be treated to remove sulfur, nitrogen, and metallic contaminants, typically using a catalytic hydrotreater (which will consume some hydrogen, but not as much as produced in the CRU). The CRU usually has a series of three or four reactors. The reforming reactor is endothermic, so the feed must be heated prior to each reactor vessel. Coke deposits slowly on the catalyst particles during the processing reaction, and this "catalyst coke" must be burned-off to reactivate the catalyst, generating CO_2, along with small amounts of CH_4 and N_2O.

There are three types of CRU based on how the regeneration of the catalyst is performed: continuous CRU, cyclic CRU, and semi-regenerative CRU. In a continuous CRU (or platformers), small quantities of the catalyst are continuously removed from a moving bed reactor system, purged, and transported to a continuously operated regeneration system. The regenerated catalyst is then recycled to the moving bed reactor. Continuous reformers generally operate at lower pressures than other reforming units, resulting in higher coke deposition rates. Cyclic CRU has an extra reactor vessel, so that one reactor vessel can be isolated from the unit for regeneration. After the first vessel is regenerated, it is brought back on-line and the second reactor vessel is then isolated and regenerated and so on until all vessels have been regenerated. Thus, in cyclic units, the CRU continues to operate and individual reactor vessels are regenerated in a cyclical process many times during a single year. In a semi-regenerative CRU, the entire reforming unit is taken off-line to regenerate the catalyst in the reactor vessels. Catalyst regeneration in a semi-regenerative CRU typically occurs once every 12 to 24 months (18 months is typical) and lasts approximately 1 to 2 weeks (U.S. EPA, 1998).

In addition to the CO_2 generated during coke burn-off, there may be some CH_4 emissions during the depressurization and purging of the reactor vessels of recycled catalyst prior to regeneration. While the CH_4 emissions from the depressurization and purging processes are expected to be negligible in most cases, natural gas (*i.e.*, CH_4) is occasionally used as the purge gas, in which case the CH_4 emissions would not be negligible.

2.2.6 Sulfur Recovery Vents

Hydrogen sulfide (H_2S) is removed from the refinery fuel gas system through the use of amine scrubbers. While the selectivity of H_2S removal is dependent on the type of amine solution used, these scrubbers also tend to extract CO_2 from the fuel gas. The concentrated sour gas is then processed in a sulfur recovery plant to convert the H_2S into elemental sulfur or sulfuric acid. CO_2 in the sour gas will pass through the sulfur recovery plant and be released in the final sulfur plant vent. Additionally, small amounts of hydrocarbons may also be present in the sour gas stream. These hydrocarbons will eventually be converted to CO_2 in the sulfur recovery plant or via tail gas incineration. The most common type of sulfur recovery plant is the Claus unit, which produces elemental sulfur. The first step in a Claus unit is a burner to convert one-third of the sour gas into sulfur dioxide (SO_2) prior to the Claus catalytic reactors. GHG emissions from the fuel fired to the Claus burner are expected to be accounted for as a combustion source. After that, the sulfur dioxide and unburned H_2S are reacted in the presence of a bauxite catalyst to produce elemental sulfur. Based on process-specific data collected in the development of emission standards for petroleum refineries, there are 195 sulfur recovery trains in the petroleum refining industry (U.S. EPA, 1998).

2.2.7 Hydrogen Plants

The most common method of producing hydrogen at a refinery is the steam methane reforming (SMR) process. Methane, other light hydrocarbons, and steam are reacted via a nickel catalyst to produce hydrogen and CO. Excess CH_4 is added and combusted to provide the heat needed by this endothermic reaction. The CO generated by the initial reaction further reacts with the steam to generate hydrogen and CO_2 (U.S. DOE, 2007). According to EIA's Refinery Capacity Report 2006 (EIA, 2006), 54 of the 150 petroleum refineries have hydrogen production capacity. CO_2 produced as a byproduct of SMR hydrogen production accounts for approximately 6 percent of GHG emissions from petroleum refineries nationwide, but can account for 25 percent of the GHG emissions from an individual refinery. Many of the hydrogen plants located at a petroleum refinery are operated by third-parties. It is unclear if the hydrogen production units reported by EIA include all hydrogen plants co-located at a refinery or only those that are directly owned and operated by the refinery.

2.2.8 Asphalt Blowing Stills

Asphalt or bitumen blowing is used for polymerizing and stabilizing asphalt to improve its weathering characteristics in the production of asphalt roofing products and certain road asphalts. Asphalt blowing involves the oxidation of asphalt flux by bubbling air through liquid asphalt flux at 260 °Celsius (C) (500 °F) for 1 to 10 hours depending on the desired characteristics of the product. The vessel used for asphalt blowing is referred to as a "blowing still." The emissions from a blowing still are primarily organic particulate with a fairly high concentration of gaseous hydrocarbon and polycyclic organic matter as well as reduced sulfur compounds. The blowing still gas also contains significant quantities of CH_4 and CO_2. The blowing still gas is commonly controlled with a wet scrubber to remove sour gas, entrained oil, particulates, and condensable organics and/or a thermal oxidizer to combust the hydrocarbons and sour gas to CO_2 and SO_2.

2.2.9 Storage Tanks

Storage tanks will generally have negligible GHG emissions except when unstabilized crude oil is stored or a methane blanket is used in the storage tank. Unstabilized crude oil is crude oil that has not been stored at atmospheric conditions for prolonged periods of time (several days to a week) prior to being received at the refinery. Most crude oil deposits also include natural gas (*i.e.*, CH_4); some of the CH_4 is dissolved in the crude oil at the pressure of the crude oil deposit. When crude oil is extracted, it is often stored temporarily at atmospheric conditions to either discharge or recover the dissolved gases. If the crude oil is transported under pressure (*e.g.*, via a pipeline) either immediately or shortly after extraction, the dissolved gases will remain in the crude oil until it reaches the refinery. The dissolved gases will be subsequently released from this "unstabilized" crude oil when the crude oil is stored at atmospheric conditions at a storage tank at the refinery.

2.2.10 Coke Calcining Units

Coke calcining units are a significant source of CO_2 emissions; however, only a few petroleum refineries have on-site coke calcining units. Coke calciners are used to burn-off sulfur, volatiles, and other impurities in the coke to produce a premium grade coke that can be used to make electrodes, anode vessels, and other products. A small fraction of the coke is consumed/pyrolyzed in the process under oxygen starved conditions; the process gas generated is then combusted in an afterburner by mixing the process gas with air in the presence of a flame. Most of the CO_2 generated from the process/afterburner system is attributable to the volatile content of the coke fed to the calciner.

2.2.11 Other Ancillary Sources

Refineries may also contain other ancillary sources of GHG emissions. Most refineries have wastewater treatment systems and some refineries have landfills. While the aerobic biodegradation of wastes is generally considered to be biogenic, anaerobic degradation of waste producing CH_4 emission is not. The high organic loads and stagnant conditions in an oil-water separator are conducive to anaerobic degradation, and the oil water separator may be a fairly significant ancillary source of CH_4 emissions. Landfills are also conducive to anaerobic degradation. Depending on the organic content of the waste material managed in a landfill, the landfill may also be a fairly significant ancillary source of CH_4 emissions.

The refinery's fuel gas system will generally contain significant concentrations of CH_4; certain process units may either generate methane or use methane and other light ends as part of the process operations (*e.g.*, SMR hydrogen production). Leaking equipment components (*e.g.*, valves, pumps, and flanges) may, therefore, be a source of CH_4 emissions. Leak detection and repair (LDAR) programs are commonly used to identify and reduce emissions from equipment components; however, most LDAR programs exclude the fuel gas system. Similar to equipment leaks, some heat exchangers may develop leaks whereby gases being cooled can leak into the cooling water. Although these leaks are not direct releases to the atmosphere, light hydrocarbons that leak into the cooling water will generally be released to the atmosphere in cooling towers (for recirculated cooling water systems) or ponds/receiving waters (in once through systems). As several heat exchangers at a refinery cool gases that contain appreciable quantities of CH_4 (e.g., a distillation column's overhead condenser), cooling towers may also be a source of CH_4 emissions. Nonetheless, CH_4 emissions from equipment leaks, either directly to the atmosphere from leaking equipment components or indirectly from cooling towers from leaking heat exchangers, are generally expected to have a minimal contribution to a typical refinery's total GHG emissions.

3.0   Summary of GHG Reduction Measures

Table 1 summarized the GHG reduction measures described in this document. Additional detail regarding these GHG reduction measures are provided in Section 4, Energy Programs and Management Systems, and Section 5, GHG Reduction Measures by Source, of this document.

Table 1. Summary of GHG Reduction Measures for the Petroleum Refining Industry

GHG Control Measure	Description	Efficiency Improvement/ GHG emission reduction	Retrofit Capital Costs ($/unit of CO_2e)	Payback time (years)	Demon-strated in Practice?	Other Factors
Energy Efficiency Programs and Systems						
Energy Efficiency Initiatives and Improvements	Benchmark GHG performance and implement energy management systems to improve energy efficiency, such as: ▪ improve process monitoring and control systems ▪ use high efficiency motors ▪ use variable speed drives ▪ optimize compressed air systems ▪ implement lighting system efficiency improvements	4-17% of electricity consumption		1-2 years	Yes	

11

GHG Control Measure	Description	Efficiency Improvement/ GHG emission reduction	Retrofit Capital Costs ($/unit of CO_2e)	Payback time (years)	Demon- strated in Practice?	Other Factors
Stationary Combustion Sources						
Steam Generating Boilers (see also ICI Boiler GHG BACT Document)						
Systems Approach to Steam Generation	Analyze steam needs and energy recovery options, including: ▪ minimize steam generation at excess pressure or volume ▪ use turbo or steam expanders when excesses are unavoidable ▪ schedule boilers based on efficiency				Yes	
Boiler Feed Water Preparation	Replace a hot lime water softener with a reverse osmosis membrane treatment system to remove hardness and reduce alkalinity of boiler feed.	70-90% reduction in blowdown steam loss; up to 10% reduction in GHG emissions		2-5 years	Yes	
Improved Process Control	Oxygen monitors and intake air flow monitors can be used to optimize the fuel/air mixture and limit excess air.	1-3% of boiler emissions		6 - 18 months	Yes	Low excess air levels may increase CO emissions.
Improved Insulation	Insulation (or improved insulation) of boilers and distribution pipes.	3-13% of boiler emissions		6 - 18 months	Yes	
Improved Maintenance	All boilers should be maintained according to a maintenance program. In particular, the burners and condensate return system should be properly adjusted and worn components replaced. Additionally, fouling on the fireside of the boiler and scaling on the waterside should be controlled.	1-10% of boiler emissions			Yes	
Recover Heat from Process Flue Gas	Flue gases throughout the refinery may have sufficient heat content to make it economical to recover the heat. Typically, this is accomplished using an economizer to preheat the boiler feed water.	2-4% of boiler emissions		2 years	Yes	
Recover Steam from Blowdown	Install a steam recover system to recover blowdown steam for low pressure steam needs (e.g., space heating and feed water preheating).	1 –3%		1 - 3 years	Yes	

GHG Control Measure	Description	Efficiency Improvement/ GHG emission reduction	Retrofit Capital Costs ($/unit of CO_2e)	Payback time (years)	Demon-strated in Practice?	Other Factors
Reduce Standby Losses	Reduce or eliminate steam production at standby by modifying the burner, combustion air supply, and boiler feedwater supply, and using automatic control systems to reduce the time needed to reach full boiler capacity.	Up to 85% reduction in standby losses (but likely a small fraction of facility total boiler emissions)		1.5 years	Yes	
Improve and Maintain Steam Traps	Implement a maintenance plan that includes regular inspection and maintenance of steam traps to prevent steam lost through malfunctioning steam traps.	1-10% of boiler emissions			Yes	
Install Steam Condensate Return Lines	Reuse of the steam condensate reduces the amount of feed water needed and reduces the amount of energy needed to produce steam since the condensate is preheated.	1- 10% of steam energy use		1-2 years	Yes	
Process Heaters						
Combustion Air Controls- Limitations on Excess air	Oxygen monitors and intake air flow monitors can be used to optimize the fuel/air mixture and limit excess air.	1-3%		6-18 months	Yes	
Heat Recovery: Air Preheater	Air preheater package consists of a compact air-to-air heat exchanger installed at grade level through which the hot stack gases from the convective section exchange heat with the incoming combustion air. If the original heater is natural draft, a retrofit requires conversion to mechanical draft.	10-15% over units with no preheat.			Yes	**May increase NOx emissions**
Combined Heat and Power						
Combined Heat and Power	Use internally generated fuels or natural gas for power (electricity) production using a gas turbine and generate steam from waste heat of combustion exhaust to achieve greater energy efficiencies			5 years	Yes	
Carbon Capture						
Oxy-combustion	Use pure oxygen in large combustion sources to reduce flue gas volumes and increase CO_2 concentrations to improve capture efficiency and costs				No	

GHG Control Measure	Description	Efficiency Improvement/ GHG emission reduction	Retrofit Capital Costs ($/unit of CO_2e)	Payback time (years)	Demon- strated in Practice?	Other Factors
Post-combustion Solvent Capture	Use solvent scrubbing, typically using monoethanolamine (MEA) as the solvent, for separation of CO_2 in post-combustion exhaust streams				Yes	
Post-combustion membranes	Use membrane technology to separate or adsorb CO2 in an exhaust stream		$55-63		No	
Fuel Gas System and Flares						
Fuel Gas System						
Compressor Selection	Use dry seal rather than wet seal compressors; use rod packing for reciprocating compressors				Yes	
Leak Detection and Repair	Use organic vapor analyzer or optical sensing technologies to identify leaks in natural gas lines, fuel gas lines, and other lines with high methane concentrations and repair the leaks as soon as possible.	80-90% of leak emissions; <0.1% refinery-wide			Yes	
Sulfur Scrubbing System	Evaluate different sulfur scrubbing technologies or solvents for energy efficiency				Yes	
Flares						
Flare Gas Recovery	Install flare gas recovery compressor system to recover flare gas to the fuel gas system			1 yr	Yes	
Proper Flare Operation	Maintain combustion efficiency of flare by controlling heating content of flare gas and steam- or air-assist rates				Yes	
Refrigerated Condensers	Use refrigerated condensers to increase product recovery and reduce excess fuel gas production				Yes	
Cracking Units						
Fluid Catalytic Cracking Units (see also: Stationary Combustion Sources; Fuel Gas System and Flares)						
Power/Waste Heat Recovery	Install or upgrade power recovery or waste heat boilers to recover latent heat from the FCCU regenerator exhaust				Yes	
High-Efficiency Regenerators	Use specially designed FCCU regenerators for high efficiency, complete combustion of catalyst coke deposits				Yes	

GHG Control Measure	Description	Efficiency Improvement/ GHG emission reduction	Retrofit Capital Costs ($/unit of CO_2e)	Payback time (years)	Demon-strated in Practice?	Other Factors
Hydrocracking Units (see also: Stationary Combustion Sources; Fuel Gas System and Flares; Hydrogen Production Units)						
Power/Waste Heat Recovery	Install or upgrade power recovery to recover power from power can be recovered from the pressure difference between the reactor and fractionation stages			2.5 years	Yes	
Hydrogen Recovery	Use hydrogen recovery compressor and back-up compressor to ensure recovery of hydrogen in process off-gas				Yes	
Coking Units						
Fluid Coking Units (see also: Stationary Combustion Sources; Fuel Gas System and Flares)						
Power/Waste Heat Recovery	Install or upgrade power recovery or waste heat boilers to recover latent heat from the fluid coking unit exhaust				Yes	
Flexicoking Units (see: Stationary Combustion Sources; Fuel Gas System and Flares)						
Delayed Coking Units (see also: Stationary Combustion Sources; Fuel Gas System and Flares)						
Steam Blowdown System	Use low back-pressure blowdown system and recycle hot blowdown system water for steam generation				Yes	
Steam Vent	Lower pressure and temperature of coke drum to 2 to 5 psig and 230°F to minimize direct venting emissions	50 to 80% reduction in direct steam vent CH_4 emissions			Yes	
Catalytic Reforming Units (see also: Stationary Combustion Sources; Fuel Gas System and Flares; Hydrogen Production Units)						
Sulfur Recovery Units						
Sulfur Recovery System Selection	Evaluate energy and CO2 intensity in selection of sulfur recovery unit and tail gas treatment system and a variety of different tail gas treatment units including Claus, SuperClaus® and EuroClaus®, SCOT, Beavon/amine, Beavon/Stretford, Cansolv®, LoCat®, and Wellman-Lord				Yes	
Hydrogen Production Units						
Hydrogen Production Optimization	Implement a comprehensive assessment of hydrogen needs and consider using additional catalytic reforming units to produce H_2				Yes	
Combustion Air and Feed/Steam Preheat	Use heat recovery systems to preheat the feed/steam and combustion air temperature	5% of total energy consumption for H_2 production			Yes	

GHG Control Measure	Description	Efficiency Improvement/ GHG emission reduction	Retrofit Capital Costs ($/unit of CO_2e)	Payback time (years)	Demon-strated in Practice?	Other Factors
Cogeneration	Use cogeneration of hydrogen and electricity: hot exhaust from a gas turbine is transferred to the reformer furnace; the reformer convection section is also used as a heat recovery steam generator (HRSG) in a cogeneration design; steam raised in the convection section can be put through either a topping or condensing turbine for additional power generation				Yes	
Hydrogen Purification	Evaluate hydrogen purification processes (i.e., pressure-swing adsorption, membrane separation, and cryogenic separation) for overall energy intensity and potential CO_2 recovery.				Yes	
Hydrotreating Units (see also: Hydrogen Production Units; Sulfur Recovery Units)						
Hydrotreater Design	Use energy efficient hydrotreater designs and new catalyst to increase sulfur removal.				Yes	
Crude Desalting and Distillation Units						
Desalter Design	Alternative designs for the desalter, such as multi-stage units and combinations of AC and DC fields, may increase efficiency and reduce energy consumption.				Yes	
Progressive Distillation Design	Progressive distillation process uses as series of distillation towers working at different temperatures to avoid superheating lighter fractions of the crude oil.	30% reduction in crude heater emissions; 5% or more refinery-wide			Yes	
Storage Tanks						
Vapor Recovery or Control for Unstabilized Crude Oil Tanks	Consider use of a vapor recovery or control system for crude oil storage tanks that receive crude oil that has been stored under pressure ("unstabilized" crude oil)	90-95% reduction in CH_4 from these tanks			Yes	
Heated Storage Tank Insulation	Insulate heated storage tanks				Yes	

4.0 Energy Programs and Management Systems

Industrial energy efficiency can be greatly enhanced by effective management of the energy use of operations and processes. U.S. EPA's ENERGY STAR Program works with hundreds of manufacturers and has seen that companies and sites with stronger energy management programs gain greater improvements in energy efficiency than those that lack procedures and management practices focused on continuous improvement of energy performance.

Energy Management Systems (EnMS) provide a framework for managing energy and promote continuous improvement. The EnMS provides the structure for an energy program and its energy team. EnMS establish assessment, planning, and evaluation procedures which are critical for actually realizing and sustaining the potential energy efficiency gains of new technologies or operational changes.

Energy management systems promote continuous improvement of energy efficiency through:
- Organizational practices and policies,
- Team development
- Planning and evaluation,
- Tracking and measurement,
- Communication and employee engagement, and
- Evaluation and corrective measures.

For nearly 10 years, the U.S. EPA's ENERGY STAR Program has promoted an energy management system approach. This approach, outlined in **Figure 5**, outlines the basic steps followed by most energy management systems approaches.

Figure 5. ENERGY STAR Guidelines for Energy Management

In recent years, interest in energy management system approaches has been growing. There are many reasons for the greater interest. These include recognition that a lack of management commitment is an important barrier to increasing energy efficiency. Lack of an effective energy team and an effective program result in poor implementation of new technologies and poor implementation of energy assessment recommendations. Poor energy management practices that fail to monitor performance do not ensure that new technologies and operating procedures will achieve their potential to improve efficiency.

Approaches to implementing energy management systems vary. EPA's ENERGY STAR Guidelines for Energy Management are available for public use on the web and provide extensive guidance (see: www.energystar.gov/guidelines). Alternatively, energy management standards are available for purchase from ANSI, ANSI MSE 2001:200 and in the future from ISO, ISO 50001.

While energy management systems can help organizations achieve greater savings through a focus on continuous improvement, they do not guarantee energy savings or CO_2 reductions alone. Combined with effective plant energy benchmarking and appropriate plant improvements, energy management systems can help achieve greater savings.

There are a variety of factors to consider when contemplating requiring certification to an Energy Management Standard established by a standards body such as ANSI or ISO. First, energy management system standards are designed to be flexible. A user of the standard is able to define the scope and boundaries of the energy management system so that single production lines, single processes, a plant or a corporation could be certified. Beyond scope, achieving certification for the first time is not based on efficiency or savings (although re-certifications at a later time could be). Finally, cost is an important factor in the standardized approach. Internal personnel time commitments, external auditor and registry costs are expensive.

From a historical perspective, few companies have pursued certification according to the ANSI energy management standards to date. One reason for this is that the elements of an energy management system can be applied without having to achieve certification which adds additional costs. The ENERGY STAR Guidelines and associated resources are widely used and adopted partly because they are available in the public domain and do not involve certification.

Overall, a systems approach to energy management is an effective strategy for encouraging energy efficiency in a facility or corporation. The focus of energy management efforts are shifted from a "projects" to a "program" approach. There are multiple pathways available with a wide range of associated costs (ENERGY STAR energy management resources are public while the standardized approaches are costly). The effectiveness of an energy management system is linked directly to the system's scope, goals and measurement and tracking. Benchmarks are the most effective measure for demonstrating the system's achievements.

4.1   Sector-Specific Plant Performance Benchmarks

Benchmarking is the process of comparing the performance of one site against itself over time or against the range of performance of the industry. Benchmarking is typically done at a whole facility or site level to capture the synergies of different technologies, operating practices, and operating conditions and typically results in a calculation of the emissions intensity of a site, which are the emissions per unit of product.

For a refinery, emissions intensity is influenced by a number of factors, including energy efficiency, fuel use, feed composition, and products. While refineries all refine crude oil to make a range of common products (gasoline, diesel, fuel oils, liquefied petroleum gases), they often vary in size and the number of processing units that are operating. For example, refineries with more simple configurations may not be able to process certain fractions into more energy-intensive products. Likewise, refineries that process heavy sour crudes may require more energy intensive processing. Benchmarking approaches have been used in the refining industry for many years to improve efficiency and productivity. The European Union evaluated and concluded that the Solomon's complexity weighted barrel approach should be used to benchmark refineries as part of their methodology for allocating emission allowances in the European Union Emissions Trading System (Ecofys, 2009).

4.2   Industry Energy Efficiency Initiatives

The U.S. EPA's ENERGY STAR Program (www.energystar.gov/industry) and U.S. DOE's Industrial Technology Program (www.energy.gov/energyefficiency) have led industry specific energy efficiency initiatives over the years. These programs have helped to create guidebooks of energy efficient technologies, profiles of industry energy use, and studies of future technologies. Some states have also led sector specific energy efficiency initiatives. Resources from these programs can help to identify technologies that may help reduce CO_2 emissions.

EPA's ENERGY STAR Program has conducted an energy efficiency improvement assessment for petroleum refineries (Worrell and Galitsky, 2005). Many of the GHG reduction measures provided in the following sections are a result of this industry-specific assessment.

4.3   Energy Efficiency Improvements in Facility Operations

4.3.1   Monitoring and Process Control Systems

Most refineries already employ some energy management systems. At existing facilities, only a limited number of processes or energy streams may be monitored and managed. Opportunities should be evaluated for expanding the coverage of monitoring systems throughout the plant. New facilities should include a comprehensive energy management program (Worrell and Galitsky, 2005).

Process control systems are available for essentially all industrial processes. These control systems are typically designed to primarily improve productivity, product quality, and efficiency of a process. However, each of these improvements will lead to increased energy efficiency as well. Process control systems also reduce downtime, maintenance costs, and processing time, and increase resource efficiency and emission control (Worrell and Galitsky, 2005).

Although specific energy savings and payback periods are highly facility-specific, the application of monitoring systems to specific industrial applications have demonstrated energy savings of 4-17 percent, and process control systems can reduce energy consumption by 2-18 percent over facilities without such systems. In general, cost and energy savings of about 5 percent can be expected through the implementation of monitoring and process control systems (Worrell and Galitsky, 2005).

Valero and AspenTech have developed a system to model and control plant-wide energy usage for refinery operations. The system was installed at a domestic refinery and is expected to reduce overall energy usage by 2-8 percent (Worrell and Galitsky, 2005).

Process control systems for the CDU have been shown to reduce energy costs by $0.05-0.12/barrel (bbl) of feed, with paybacks of less than 6 months. Another CDU control system reduced energy consumption and flaring and increased throughput, resulting in a payback of about 1 year. In Portugal, a refinery installed advanced CDU controls and realized a 3-6 percent increase in throughput. The payback period for this control system was 3 months (Worrell and Galitsky, 2005).

Process control systems for FCCU are supplied by several companies. Cost savings range from $0.02-0 40.bbl of feed with paybacks ranging from 6-18 months. At one refinery, an existing FCCU control system was updated at a 65,000 bpd unit and a cost savings of $0.05/bbl of feed was realized. A refinery in Italy installed a control system on a FCCU and reduced cost by $0.10/bbl of feed with a payback of less than 1 year. (Worrell Galitsky, 2005)

In South Africa, a refinery installed a multivariable predictive control system on a hydrotreater. Hydrogen consumption was reduced by 12 percent and the fuel consumption of the heater was reduced by 18 percent. Improved yield of gasoline and diesel were also realized. The payback period was 2 months (Worrell and Galitsky, 2005).

4.3.2 High Efficiency Motors

Electric motors are used throughout the refinery for such applications as pumps, air compressors, fans, and other applications. Pumps, compressors and fans account for 70 to 80 percent of the total electricity usage at the refinery (Worrell and Galitsky, 2005). As such, a systems approach to energy efficiency should be considered for all motor systems (motors, drives, pumps, fans, compressors, controls). An evaluation of energy supply and energy demand could be performed to optimize overall performance. A systems approach includes a motor management plan that considers at least the following factors (Worrell and Galitsky, 2008):

- Strategic motor selection
- Maintenance
- Proper size
- Adjustable speed drives
- Power factor correction
- Minimize voltage unbalances

Pumps are the single largest electricity user at a refinery, accounting for about half of the total energy usage. One study estimated that 20 percent of the energy consumed by pump motors could be saved through equipment or control system changes. Implementation of maintenance programs for pump motors can reduce electricity use by 2-7 percent, with payback periods less than 1 year (Worrell and Galitsky, 2005).

Motor management plans and other efficiency improvements can be implemented at existing facilities and should be considered in the design of new construction. At existing facilities, replacing older motors with high efficiency motors are typically cost-effective when a motor needs replacement, but may not be economical when the old motor is still operational. Payback periods from energy savings are typically less than 1 year (Worrell and Galitsky, 2005).

4.3.3 Variable Speed Drives

Energy use on centrifugal systems such as pumps, fans, and compressors is approximately proportional to the cube of the flow rate. Therefore, small reductions in the flow may result in large energy savings. The use of variable speed drives can better match speed to load requirements of the motors. The installation of variable speed drives at new facilities can result in payback periods of just over 1 year (Worrell and Galitsky, 2005).

4.3.4 Optimization of Compressed Air Systems

Compressed air systems provide compressed air that is used throughout the refinery. Although the total energy used by compressed air systems is small compared to the facility as a whole, there are opportunities for efficiency improvements that will save energy. Efficiency improvements are primarily obtained by implementing a comprehensive maintenance plan for the compressed air systems. Worrell and Galitsky (2005, 2008) listed the following elements of a proper maintenance plan:

- Keep the surfaces of the compressor and intercooling surfaces clean
- Keep motors properly lubricated and cleaned
- Inspect drain traps
- Maintain the coolers
- Check belts for wear
- Replace air lubricant separators as recommended
- Check water cooling systems

In addition to the maintenance plan, reducing leaks in the system can reduce energy consumption by 20 percent. Reducing the air inlet temperature will reduce energy usage, and routing the air intake to outside the building can have a payback in 2-5 years. Control systems can reduce energy consumption by as much as 12 percent. Properly sized pipes can reduce energy consumption by 3 percent. Since as much as 93 percent of the electrical energy used by air compressor systems is lost as heat, recovery of this heat can be used for space heating, water heating, and similar applications (Worrell and Galitsky, 2005, 2008).

Air compressor system maintenance plans and other efficiency improvements can be implemented at existing facilities and should be considered in the design of new construction.

4.3.5 Lighting System Efficiency Improvements

Similar to air compressor systems, the energy used for lighting at a petroleum refinery facilities represent a small portion of the overall energy usage. However, there are opportunities for cost-effective energy efficiency improvements. Automated lighting controls that shut off lights when not needed may have payback periods of less than 2 years. Replacing T-12 lights with T-8 lights can reduce energy use by half, as can replacing mercury lights with metal halide or high pressure sodium lights. Substituting electronic ballasts for magnetic ballasts can reduce energy consumption by 12-25 percent (Worrell and Galitsky, 2005, 2008).

Lighting system improvements can be implemented at existing facilities and should be considered in the design of new construction.

5.0   GHG Reduction Measures by Source

5.1   *Stationary Combustion Sources*

5.1.1   Steam Generating Boilers

According to Worrell and Galitsky (2005), approximately 30 percent of onsite energy use at domestic refineries is used in the form of steam generated by boilers, cogeneration, or waste heat recovery from process units. The U.S. DOE estimated steam accounts for 38 percent of a refinery's energy needs (U.S. DOE, 2002). However, off-site purchases of steam represent only 3 to 5 percent of the total energy consumption at petroleum refineries nationwide (EIA, 2009). Given that steam accounts for 30 to 38 percent of a refinery's energy needs, it is evident that most refineries produce their own steam. As such, steam generation and distribution makes a significant contribution to a petroleum refinery's energy needs, and subsequently its on-site GHG emissions.

5.1.1.1 *Systems Approach to Steam Generation*

A thorough analysis of steam needs and energy recovery opportunities could be conducted to make the steam generation process as efficient as possible. For example, the analysis should assure that steam is not generated at pressures or volumes larger than what is needed. In those situations where the steam generation has limited adjustability, the excess energy in the steam should be recovered using a turbo expander or steam expansion turbine. Another option is to operate multiple boilers that are regulated according to steam demands. One refinery that implemented a program including scheduling of boilers on the basis of efficiency and minimizing losses in the turbines resulted in $5.4 million in energy savings (Worrell and Galitsky, 2005).

5.1.1.2 *Boiler Feed Water Preparation*

Boiler feed water is typically pre-treated to remove contaminates that foul the boiler. A refinery in Utah replaced a hot lime water softener with a reverse osmosis membrane treatment system to remove hardness and reduce alkalinity. Blowdown was reduced from 13.3 percent to 1.5 percent of steam produced. Additionally, reductions were seen in chemical usage, maintenance, and waste disposal costs. The initial investment of the membrane system was $350,000 and annual savings of $200,000 were realized (Worrell and Galitsky, 2005).

5.1.1.3 *Improved Process Control*

Boilers are operated with a certain amount of excess air to reduce emissions and for safety considerations. However, too much excess air may lead to inefficient combustion, and energy must be used to heat the excess air. Oxygen monitors and intake air flow monitors can be used to optimize the fuel/air mixture. Payback for such systems is typically about 0.6 years (Worrell and Galitsky, 2005).

5.1.1.4 Improved Insulation

The insulation of older boilers may be in poor condition, and the material itself may not insulate as well as newer materials. Replacing the insulation combined with improved controls can reduce energy requirements by 6-26 percent. Insulation on steam distribution systems should also be evaluated. Improving the insulation on the distribution pipes at existing facilities may reduce energy usage by 3-13 percent, with an average payback period of 1.1 years (Worrell and Galitsky, 2005).

5.1.1.5 Improved Maintenance

All boilers should be maintained according to a maintenance program. In particular, the burners and condensate return system should be properly adjusted and worn components replaced. Average energy savings of about 10 percent can be realized over a system without regular maintenance. Additionally, fouling on the fireside of the boiler and scaling on the waterside should be controlled (Worrell and Galitsky, 2005).

5.1.1.6 Recover Heat from Boiler Flue Gas

Flue gasses throughout the refinery may have sufficient heat content to make it economical to recover the heat. Typically, this is accomplished using an economizer to preheat the boiler feed water. One percent of fuel use can be saved for every 25 °C reduction in flue gas temperature. In some situations, the payback for installing an economizer is about 2 years (Worrell and Galitsky, 2005).

5.1.1.7 Recover Steam from Blowdown

The pressure drop during blowdown may produce substantial quantities of low grade steam that is suitable for space heating and feed water preheating. For boilers below 100 MMBtu/yr, fuel use can be reduced by about 1.3 percent, and payback may range from 1-2.7 years. A chemical plant installed a steam recover system to recover all of the blowdown steam from one process and realized energy savings of 2.8 percent (Worrell and Galitsky, 2005).

5.1.1.8 Reduce Standby Losses

It is common practice at most refineries to maintain at least one boiler on standby for emergency use. Steam production at standby can be virtually eliminated by modifying the burner, combustion air supply, and boiler feed water supply. Additionally, automatic control systems can reduce the time needed to reach full capacity of the boiler to a few minutes. These measures can reduce the energy consumption of the standby boiler by as much as 85 percent Worrell and Galitsky, 2005).

These measures were applied to a small 40 tonnes/hr steam boiler at an ammonia plant, resulting in energy savings of 54 TBtu/yr with a capital investment of about $270,000 (1999$). The payback period was 1.5 years (Worrell and Galitsky, 2005).

5.1.1.9 Improve and Maintain Steam Traps

Significant amounts of steam may be lost through malfunctioning steam traps. A maintenance plan that includes regular inspection and maintenance can reduce boiler energy usage by up to 10 percent (Worrell and Galitsky, 2005).

5.1.1.10 Install Steam Condensate Return Lines

Reuse of the steam condensate reduces the amount of feed water needed and reduces the amount of energy needed to produce steam since the condensate is preheated. The costs savings can justify the cost of the condensate return lines. Estimates of energy savings are as high as 10 percent, with a payback period of 1.1 years for facilities with no or insufficient condensate return systems (Worrell and Galitsky, 2005).

5.1.2 Process Heaters

5.1.2.1 Draft Control

Excessive combustion air reduces the efficiency of process heater burners. At one domestic refinery, a control system was installed on three CDU furnaces to maintain excess air at 1 percent rather than the previous 3-4 percent. Energy usage of the burners was reduced by 3-6 percent and nitrogen oxide (NO_x) emissions were reduced by 10-25 percent. The cost savings due to reduced energy requirements was $340,000. Regular maintenance of the draft air intake systems can reduce energy usage and may result in payback periods of about 2 months (Worrell and Galitsky, 2005). Draft control is applicable to new or existing process heaters, and is cost-effective for a wide range of process heaters (20 to 30 MMBtu/hr or greater).

5.1.2.2 Air Preheating

The flue gases of the furnace can be used to preheat the combustion air. Every 35 °F drop in exit flue gas temperature increases the thermal efficiency of the furnace by 1 percent. The resulting fuel savings can range from 8-18 percent, and may be economically attractive when the flue gas temperature is above 650 °F and the heater size is 50 MMBtu/hr or more. Payback periods are typically on the order of 2.5 years. One refinery in the United Kingdom installed a combustion air preheater on a vacuum distillation unit (VDU) and reduced energy costs by $109,000/yr. The payback period was 2.2 years (Worrell and Galitsky, 2005). Air preheating would require natural draft system to be converted to a forced draft system requiring installation of fans, which would increase electricity consumption and typically increase NO_X emissions. Consequently, several factors, including process heater size and draft type as well as secondary impacts, need to be considered retrofitting existing process heaters. Air preheating is often much more economical and effective when considered in the design of a new process heater.

5.1.3 Combined Heat and Power (CHP)

The large steam requirements for refining operations and the continuous operations make refineries excellent candidates for combined heat and power (CHP) generation. Refineries represent one of the largest industry sources of CHP today with 103 active CHP plants with an electric generation capacity of 14.6 gigawatts (ICF, 2010). Currently, about 60-70 percent of the 137 refineries operating at the beginning of 2010 use CHP (ICF International, 2010; EIA, 2009).

About 75 percent of the refinery CHP capacity comes from natural gas-fired combined cycle power plants consisting of large combustion turbines with heat recovery steam generators (HRSG) producing power and steam. A portion of the steam produced is used to generate more power in back pressure steam turbines. These plants meet the facility steam loads but often produce much more power than is needed by the facility itself, and, therefore, export power to the electric grid. The next most common type of CHP system is a combustion turbine with heat recovery. These systems make up about 11 percent of the existing refinery CHP capacity. Again, these systems are fueled mostly with natural gas, but internally generated fuels (*i.e.,* refinery fuel gas) are also used. Most of the remaining system CHP capacity is boilers producing high pressure steam that run through a back-pressure steam turbine to produce power and lower pressure steam for process use. These systems generally do not use natural gas but, instead, are fired with a variety of internally generated fuels, waste fuels, and even coal.

While CHP systems are already in use at the majority of domestic refineries, there are significant remaining opportunities to add CHP-based on evaluation of steam requirements met by boilers and by CHP (Worrell and Galitsky, 2005). In addition, there are opportunities to repower existing CHP plants making them larger and more efficient by adding newer, more efficient combustion turbines and by converting existing simple cycle plants to combined cycle operation by adding steam turbines for additional power. Additionally, as refineries install flare gas recovery systems, they may need to install CHP systems to provide a productive source for utilizing the recovered fuel gas. There may be no direct CO_2 reductions at refineries from this technology, but indirect reductions from displacing grid power. The level of reduction is a function of the CO_2 intensity of the displaced external power production.

CHP systems require a fairly substantial investment ($1,000-2,500/kilowatt (kW)); however, the economics of CHP operation at refineries is generally very attractive. One refinery installed a 34 megawatt (MW) cogeneration unit in 1990 that consisted of two gas turbines and two heat recovery steam boilers. All facility electricity needs are met by the unit, and occasionally excess electricity is exported to the grid. Cost savings resulting from the onsite production of electricity were about $55,000/day. CHP can also be economical for small refineries. One study for an asphalt refinery showed that a 6.5 MW gas turbine CHP unit would reduce energy costs by $3.8 million/yr with a payback period of 2.5 years (Worrell and Galitsky, 2005).

5.1.4 Carbon Capture

The post-combustion technologies listed below are generally end-of-pipe measures. It should be noted that petroleum refineries emit CO_2 from a number of different process, and the exhaust stacks for these emission points are numerous and scattered across the facility. The consideration of CO_2 capture and control at a refinery would likely be limited to the larger CO_2 emitting stacks, such as the FCCU, the fluid coking unit, the hydrogen plant, and large boilers or process heaters.

5.1.4.1 Oxy-Combustion

Oxy-combustion is the process of burning a fuel in the presence of pure or nearly pure oxygen instead of air. Fuel requirements are reduced because there is no nitrogen component to be heated, and the resulting flue gas volumes are significantly reduced (Barker, 2009).

The process uses an air separation unit to remove the nitrogen component from air. The oxygen-rich stream is then fed to the combustion unit so the resulting exhaust gas contains a higher concentration of CO_2, as much as 80 percent. A portion of the exhaust stream is discharged to a CO_2 separation, purification, and compression facility. The higher concentration of CO_2 in the flue gas directly impacts size of the adsorber (or other separation technique), and the power requirements for CO_2 compression. This technology is still in the research stage. The

Petroleum Environmental Research Forum (PERF) is focusing on large refinery combustion sources, particularly the FCCU and crude oil process heaters.

5.1.4.2 Post-Combustion Solvent Capture and Stripping

Post-combustion capture using solvent scrubbing, typically using monoethanolamine (MEA) as the solvent, is a commercially mature technology. Solvent scrubbing has been used in the chemical industry for separation of CO_2 in exhaust streams (Bosoaga, 2009).

5.1.4.3 Post-Combustion Membranes

Membrane technology may be used to separate or adsorb CO_2 in an exhaust stream. It has been estimated that 80 percent of the CO_2 could be captured using this technology. The captured CO_2 would then be purified and compressed for transport. Initial projections of specific costs range from \$55-63/tonne CO_2 avoided for cement manufacturing. The current state of this technology is primarily the research stage, with industrial application at least 10 years away. Positive aspects of membrane systems include very low maintenance (no regeneration required) (ECRA, 2009).

5.2 Fuel Gas Systems and Flares

5.2.1 Fuel Gas Systems

Many process units at the refinery, particularly atmospheric crude oil distillation, catalytic cracking, catalytic hydrocracking, thermal cracking, and coking processes, produce fuel gas that is commonly recovered for use in process heaters and boilers throughout the refinery. Typically a compressor is needed to recover the fuel gas at the fuel gas producing unit. The fuel gas generally needs to be treated to remove H_2S using amine scrubber systems. The remainder of the fuel gas system consists of piping and mix drums to transport the fuel gas to the various combustion sources at the refinery. Rather than repeating the GHG reduction measures for each potential fuel gas producing units, the GHG reduction measures for the fuel gas system are summarized here.

5.2.1.1 Compressor Selection

Different types of compressors have different propensities to leak. Based on emission factors for natural gas compressors, reciprocating compressors generally have approximately one-half the fugitive emissions of centrifugal compressors (U.S. EPA, 1999). Rod packing (e.g., Static-Pac) can be used to reduce fugitive emissions from reciprocating compressors, and dry seal centrifugal compressors have lower emissions (i.e., are less likely to leak) than those with wet seals (U.S. EPA, 1999). Thus, the projected methane emissions from fuel gas compressors could be considered in the selection of the type of compressor and fugitive controls used.

5.2.1.2 Leak Detection and Repair (LDAR)

LDAR programs have been used to reduce emissions of volatile organic compounds (VOC) from petroleum refineries for years. However, CH_4 is not a VOC, so current regulations do not generally require LDAR for refinery fuel gas systems or other high CH_4-containing gas streams. Leaks can be detected using organic vapor analyzers or specially designed cameras. LDAR programs commonly achieve emission reduction efficiencies of 80 to 90 percent; however, CH_4 emissions from leaking equipment components is expected to have a minimal contribution to the refinery's total GHG emissions.

5.2.1.3 Selection of Fuel Gas Sulfur Scrubbing System

Hydrogen sulfide in fuel gas is commonly removed by amine scrubbing. The scrubbing solution is typically regenerated by heating the scrubbing solution in a stripping column, typically using steam. The regeneration process can use significant energy, and the energy intensity (impacting CO_2 emissions) of the different processes should be considered (in conjunction with the sulfur scrubbing efficiencies) in selecting scrubbing technology. Some fuel gas, such as fuel gas produced by coking units, contain a significant quantity of other reduced sulfur compounds, such as methyl mercaptan and carbon disulfide, that are not removed by conventional amine scrubbing. The impact of these other reduced sulfur compounds on the resulting sulfur dioxide (SO_2) emissions from process heaters and other fuel gas combustion devices using coker-produced fuel gas should be considered for both energy efficiency (for GHG emission reductions) and total sulfur removal efficiency (for SO_2 emission reductions). Alternatives to conventional amine scrubbing (which uses dimethylethylamine, DMEA), include the use of proprietary scrubbing systems, such as FLEXSORB®, Selexol®, and Rectisol®, as well as using a mixture of solvents as in the Sulfinol process, additional conversion of sulfur compounds to H2S prior to scrubbing, or using a direct fuel gas scrubbing/sulfur recovery technology like LoCat® or caustic scrubbers.

CO_2 is also removed by amine scrubbing; however, this will not really impact the CO_2 emissions from the plant unless sulfur recovery occurs offsite because the CO_2 will be emitted either from the combustion unit receiving the fuel gas or from the sulfur recovery unit receiving the sour gas from the amine scrubbers. Therefore, the CO_2 scrubbing efficiency of the amine scrubbers is not important; however, some light hydrocarbons may also dissolve in the amine solution and subsequently sent to the sulfur recovery plant in the sour gas stream. Most hydrocarbons in the sour gas will eventually be oxidized in the sulfur recovery plant, so entrainment of hydrocarbons does lead to additional CO_2 emissions. Therefore, scrubbing systems could be evaluated based on their sulfur removal efficiency, energy efficiency, and ability to not entrain hydrocarbons. Note that higher sulfur removal efficiencies may have an energy penalty (*i.e.*, requiring more regeneration steam per pound of treated fuel gas), so a holistic analysis is needed when selecting the sulfur scrubbing system.

5.2.2 Flares

5.2.2.1 Flare Gas Recovery

Flaring can be reduced by installation of commercially available recovery systems, including recovery compressors and collection and storage tanks. Such systems have been installed at a number of domestic refineries. At one 65,000 bpd facility in Arkansas, two flare gas recovery systems were installed that reduced flaring almost completely. This facility will use flaring only in emergencies when the amount of flare gas exceeds the capacity of the recovery system. The recovered gas is compressed and used in the refinery's fuel system. The payback period for flare gas recovery systems may be as little as 1 year (Worrell and Galitsky, 2005). Similar flare gas recovery projects have been reported in the literature (John Zinc Co, 2006; Envirocomb Limited, 2006; Peterson *et al.*, 2007; U.S. DOE, 2005), reducing flaring by approximately 95 percent. Based on emission inventory presented by Lucas (2008), nationwide CO_2 emissions from flaring at petroleum refineries were estimated to be 5 million metric tons. Provided that the recovered fuel can off-set natural gas purchases, flare gas recovery is generally cost-effective for recovering routine flows of flare gas exceeding 20 MMBtu/hr (approximately

0.5 to 1-million scf per day, depending on heat content of flare gas). Based on these estimates, flare gas recovery could reduce nationwide CO_2 emissions from flares by 3-million metric tons. The cost-effectiveness of flare gas recovery is highly dependent on the heating value of the flare gas to be recovered and the price of natural gas. For refineries that may have excess fuel gas, a flare gas recovery system may also need to include a combined heat and power unit to productively use the recovered flare gas as described in Section 5.1.

5.2.2.2 Proper Flare Operation

Poor flare combustion efficiencies generally lead to higher methane emissions and therefore higher overall GHG emissions due to the higher global warming potential (GWP) of methane. Poor flare combustion efficiencies can occur at very low flare rates with high crosswinds, at very high flow rates (*i.e.*, high flare exit velocities), when flaring gas with low heat content, and excessive steam-to-gas mass flows. Installing flow meters and gas composition monitors on the flare gas lines and having automated steam rate controls allows for improved flare gas combustion control, and minimizes periods of poor flare combustion efficiencies.

5.2.2.3 Refrigerated Condensers for Process Unit Distillation Columns

For refineries that are rich in fuel gas, an alternative to a flare gas recovery system and CHP unit may be the use of a refrigerated condenser for distillation column overheads. Product recovery may be limited by the temperature of the distillation unit overhead condenser, causing more gas to be sent to the refinery fuel gas system and/or flare. The recovery temperature can be reduced by installing a waste heat driven refrigeration plant. A refinery in Colorado installed such a system in 1997 on a catalytic reforming unit distillation column and was able to recover 65,000 bbl/yr of LPG that was previously flared or used as a fuel. The payback of the system was about 1.5 years (Worrell and Galitsky, 2005).

5.3   Cracking Units

5.3.1   Catalytic Cracking Units

5.3.1.1 Power/Waste Heat Recovery

The most likely candidate for energy recovery at a refinery is the FCCU, although recovery may also be obtained from the hydrocracker and any other process that operates at elevated pressure or temperature. Most facilities currently employ a waste heat boiler and/or a power recovery turbine or turbo expander to recover energy from the FCCU catalyst regenerator exhaust. Existing energy recovery units should be evaluated for potential upgrading. One refinery replaced an older recovery turbine and saw a power savings of 22 MW and will export 4 MW to the power grid. Another facility replaced a turbo expander and realized a savings of 18 TBtu/yr (Worrell and Galitsky, 2005).

5.3.1.2 High-Efficiency Regenerators

High efficiency regenerators are specially designed to allow complete combustion of coke deposits without the need for a post-combustion device reducing auxiliary fuel combustion associated with a CO boiler.

5.3.1.3 Additional Considerations

Catalytic cracking units are significant fuel gas producers. As such, an FCCU can significantly alter the fuel gas balance of the refinery and may cause the refinery to be fuel gas rich (produce more fuel gas than it consumes) or increase the frequency of flare gas system overpressurization to the flare. GHG measures for fuel gas systems could be considered. Flare gas recovery for the impacted flare(s) could also be considered. Also, an FCCU will have a process heater to heat the feed, so GHG reduction measures for process heaters may also need to be considered. Finally, as FCCUs are one of the largest single CO_2 emission sources at a refinery, carbon capture techniques (Section 5.1.4) could be considered.

5.3.2 Hydrocracking Units

5.3.2.1 Power/Waste Heat Recovery

For hydrocracker units, power can be recovered from the pressure difference between the reactor and fractionation stages. In 1993, one refinery in the Netherlands installed a 910 kW power recovery turbine to replace the throttle at its hydrocracker unit at a cost of $1.2 million (1993$). The turbine produced about 7.3 million kilowatt hour per year (kWh/yr) and had a payback period of 2.5 years (Worrell and Galitsky, 2005).

5.3.2.2 Hydrogen Recovery

The hydrocracking unit is a significant consumer of hydrogen. Therefore, it is likely that a hydrocracking unit will significantly impact hydrogen production rates at the refinery (if the hydrogen production unit is captive to the refinery, i.e., under common ownership or control). The off-gas stream of the hydrocracker contains a significant amount of hydrogen, which is typically compressed, recovered, and recycled to the hydrocracking unit. When the recovery compressor fails or is taken off-line for maintenance, this high hydrogen gas stream is typically flared. A back-up recovery compressor could be considered for this high hydrogen stream. Although the flaring of hydrogen does not directly produce GHG, if natural gas is added to supplement the heating value of the flare gas, then flaring of the gas stream generates GHG. More importantly, the recovery of the hydrogen in this off-gas directly impacts the net quantity of new hydrogen that has to be produced for the unit. As hydrogen production has a large CO_2 intensity, continuous recovery of this high hydrogen gas stream can result in significant CO_2 emission reductions. At one Texas refinery, replacement of the hydrogen gas stream recovery compressor took 6 months, over which period approximately 7,000 tonnes of H_2 was flared, which corresponds to 63,000 to 70,000 tonnes of CO_2 emissions from additional hydrogen production. Considering the annualized capital cost of a back-up recovery compressor, the costs associated with the GHG emission reductions in this instance would be approximately $20 per tonne of CO_2 reduced.

5.3.2.3 Additional Considerations

Hydrocracking units produce fuel gas. As such, GHG measures for fuel gas systems are likely applicable for hydrocracking units. Additionally, flare gas recovery for the impacted flare(s) could be considered. The hydrocracking unit will have a process heater to heat the feed, so GHG reduction measures for process heaters may also need to be considered.

5.4 Coking Units

5.4.1 Fluid Coking Units

5.4.1.1 Power/Waste Heat Recovery

The fluid coking unit is an excellent candidate for energy recovery at a refinery. A CO boiler is used to combust the high CO off-gas from the fluid coking unit. Steam generation and/or a power recovery turbine or turbo expander could be used to recover energy from the CO boiler and its exhaust stream. Existing energy recovery units could be evaluated for potential upgrading.

5.4.1.2 Additional Considerations

Fluid coking units are significant fuel gas producers; GHG measures for fuel gas systems should be considered. Flare gas recovery for the impacted flare(s) could also be considered. The fluid coking unit will have a process heater to preheat the feed. Heat recovery systems could be considered for feed preheat; GHG reduction measures for process heaters may also need to be considered. Finally, as fluid coking units are one of the largest single CO_2 emission sources at a refinery, carbon capture techniques (Section 5.1.4) could be considered.

5.4.2 Flexicoking Units

Flexicoking coking units primarily produce a low-heating value fuel gas. Heat recovery from the produced gas stream should be used to preheat feed or to generate steam. The low-heating value fuel gas is typically combusted in specialized boilers and the GHG reduction measures for boilers could be reviewed. Also, flare gas recovery for the impacted flares and GHG reduction measures for process heaters may also need to be considered.

5.4.3 Delayed Coking Units

5.4.3.1 Steam Blowdown System

Delayed coking units use steam to purge and cool coke drums that have been filled with coke as the first step in the decoking process. A closed blowdown system for this steam purge controls both VOCs and methane. The steam to the blowdown system from a DCU will contain significant concentrations of methane and light VOCs. These systems could be enclosed to prevent fugitive emissions from the offgas or collected water streams. The noncondensibles from the blowdown system could be either recovered or directly sent to a combustion device, preferably a process heater or boiler rather than a flare to recover the energy value of the light hydrocarbons. Note that the sulfur content of this gas may prevent its direct combustion without treatment to remove sulfur.

As noted previously in Section 5.1.1.7 (regarding steam generating boilers), the blowdown system could be designed to operate at low pressures, so the DCU can continue to purge to the blowdown system rather than to atmosphere for extended periods. Also, a recovery unit to recycle hot blowdown system water for steam generation should be evaluated to improve the energy efficiency associated with the DCU's steam requirements.

5.4.3.2 Steam Vent

The DCU "steam vent" is potentially a significant emission source of both methane and VOCs. While not completely understood, the emissions from this vent are expected to increase based on the coke drum vessel pressure and the average temperature when the steam off-gas is first diverted to the atmosphere at (rather than to the blowdown system) at the end of the coke drum purge and cooling cycle. Generally, cycle times of 16 to 20 hours are needed to purge, cool, and drain the coke drum vessels, cut the coke out, and preheat the vessel prior to receiving feed. In efforts to increase throughput of the unit, reduced cycle times are used, but this generally requires depressurization of the coke drum at higher temperatures and pressures leading to higher emissions. While larger coke drums may have slightly higher emissions than smaller coke drums, the temperature of the coke drum when the drum is first vented to atmosphere will have a more significant impact on the volume of gas vented to the atmosphere than does the size (volume) of the coke drum. Cycle times of less than 16 hours are an indicator that the purging/quench cycles may be too short, leading to excessive and unnecessary VOC and CH_4 emissions. 40 CFR Part 60 subpart Ja requires new DCU to not vent to the atmosphere until a vessel pressure of 5 psig or less is reached. At this pressure, the equilibrium coke bed temperature should be approximately 230°F. However, as the vessel will be continuously purging to the blowdown system, the bed temperature may be significantly higher even though the pressure of the vessel is below 5 pounds per square inch gauge (psig) depending on the cycle time. A DCU could be designed to allow depressurization to very low pressures (*e.g.,* 2 psig) prior to having to go to atmosphere (which will impact the blowdown system design) to allow flexibility in operation. Analysis of the CH_4 and VOC emissions at different temperatures and pressures could be conducted to determine operational parameters for the DCU depressurization/steam vent.

5.4.3.2 Additional Considerations

Delayed coking units are significant fuel gas producers. As such, GHG measures for fuel gas systems and flares could be considered. The fluid coking unit will have a process heater to preheat the feed. Heat recovery systems could be considered for feed preheat; GHG reduction measures for process heaters may also need to be considered.

5.5   *Catalytic Reforming Units*

The catalytic reforming unit is a net producer of hydrogen, so it can be considered as a means to produce hydrogen needed for other processes at the petroleum refineries; more detailed discussion of this is provided in Section 5.7. The reforming reaction is endothermic, so the catalytic reforming unit has large process heaters to maintain the reaction; GHG reduction measures for the process heaters could be considered. The catalytic reforming unit will also produce fuel gas so that GHG reduction measures for fuel gas systems and flares could be considered.

5.6   *Sulfur Recovery Units*

Nearly all refineries use the Claus-based sulfur recovery units, although some small refineries use LoCat™ system. There are, however, some variations on the traditional Claus system (*e.g.,* SuperClaus® and EuroClaus®) and a variety of different tail gas treatment units that are used in conjunction with the Claus sulfur recovery systems (*e.g.,* SCOT, Beavon/amine; Beavon/Stretford; Cansolv®, LoCat®, and Wellman-Lord). The energy and CO_2 intensities of these different systems could be evaluated (in conjunction with their sulfur recovery efficiencies) for sulfur recovery systems.

5.7   *Hydrogen Production Units*

Hydrotreating and hydrocracking units consume hydrogen. Hydrogen is produced as a by-product in catalytic reforming units. Hydrogen may also be produced specifically in captive or merchant hydrogen production units, which typically use steam methane reforming (SMR) techniques. Due to the importance of hydrogen for key processes and the interlinking of processes, a facility-wide hydrogen assessment could be performed to assess energy and GHG improvements that can be made. This assessment could include an assessment of whether additional catalytic reforming capacity can meet the hydrogen needs. Although both catalytic reforming and SMR are endothermic and require significant heat input, catalytic reformers produce high octane reformate (cyclic and aromatic hydrocarbons) rather than CO_2 as a result of the reforming reactions. Therefore, catalytic reforming provides a less CO_2-intensive means of producing hydrogen as compared to SMR hydrogen production. However, there is a limited quantity of naphtha and a limited need for reformate, so catalytic reforming may not be a viable option for meeting all of the hydrogen demands of the refinery.

If a hydrogen production unit is necessary, SMR technology appears to be the most effective means of producing additional hydrogen at this time. The following technologies could be considered for SMR hydrogen production units.

5.7.1 Combustion Air and Feed/Steam Preheat

Heat recovery systems can be used to preheat the feed/steam and combustion air temperature. If steam export needs to be minimized, an increase in the combustion air and feed/steam temperature through the convective section of the reformer is an option that can reduce fuel usage by 42 percent and steam export by 36 percent, and result in a total energy savings of 5 percent compared to a typical SMR (ARCADIS, 2008).

5.7.2 Cogeneration

Cogeneration of hydrogen and electricity can be a major enhancement of energy utilization and can be applied with SMR. Hot exhaust from a gas turbine is transferred to the reformer furnace. This hot exhaust at ~540 °C still contains ~13-percent oxygen and can serve as combustion air to the reformer. Since this stream is hot, fuel consumption in the furnace is reduced. The reformer convection section is also used as a HRSG in a cogeneration design. Steam raised in the convection section can be put through either a topping or condensing turbine for additional power generation. This technology is owned by Air Products and Technip, and has been applied at six hydrogen/cogeneration facilities for refineries (ARCADIS, 2008).

5.7.3 Hydrogen Purification

There are three main hydrogen purification processes. These are pressure-swing adsorption, membrane separation, and cryogenic separation. The selection of the purification method depends, to some extent, on the purity of the hydrogen produced. Pressure-swing adsorption provides the highest purity of hydrogen (99.9+ percent), but all of these purification methods can produce 95 percent or higher purity hydrogen stream. When lower purity (i.e., 95%) hydrogen gas is acceptable for the refinery applications, then any of the purification methods are technically viable. In such cases, the energy and CO_2 intensity of the various purification techniques could be considered. The purification technique also impacts the ease by which CO_2 recovery and capture can be used. See also the carbon capture techniques in Section 5.1.4.

5.8 *Hydrotreating Units*

A number of alternative hydrotreater designs are being developed to improve efficiency. New catalysts are being developed to increase sulfur removal, and reactors are being designed to integrate process steps. While many of these designs have not yet been proven in production, others such as oxidative desulfurization and the S Zorb process have been demonstrated at refineries. The design of both modifications and new facilities could consider the current state of the art (Worrell and Galitsky, 2005). Hydrotreaters consume hydrogen, so new hydrotreating units may also increase hydrogen production at the facility (see Section 5.7). Hydrotreaters also produce sour gas so the GHG reduction options discussed for sulfur scrubbing technologies (Section 5.2.1.3) and sulfur recovery units (Section 5.6) could be considered.

5.9   *Crude Desalting and Distillation Units*

Before entering the distillation tower, crude undergoes desalting at temperature ranging from 240 to 330 °F. Following desalting, crude enters a series of exchangers, known as preheat train to raise the temperature of the crude oil to approximately 500 °F. A direct-fired furnace is typically then used to heat the crude oil to between 650 and 750 °F before the crude oil is transferred to the flash zone of the tower. The crude oil furnaces are among the largest process heaters at the refinery; GHG reduction measures for these furnaces could be considered. Also, as the crude distillation unit employs among the largest process heaters at a refinery, carbon capture techniques (Section 5.1.4) could be considered. Additional GHG reduction measures are described below.

5.9.1 Desalter Design

Alternative designs for the desalter, such as multi-stage units and combinations of AC and DC fields, may increase efficiency and reduce energy consumption (Worrell and Galitsky, 2005).

5.9.2 Progressive Distillation Design

In the conventional scheme, all the crude feed is heated to a high temperature through the furnace prior to entering the atmospheric tower. Some lighter components of crude are superheated in the furnace, resulting in an irreversible energy waste. The progressive distillation process uses a series of distillation towers working at different temperatures (see **Figure 6**). The advantage of progressive distillation is that it avoids superheating of light fractions to temperatures higher than strictly necessary for their separation. The energy savings with progressive distillation has been reported to be approximately 30 percent (ARCADIS, 2008). Crude heaters account for approximately 25 percent of process combustion CO_2 emissions (Coburn, 2007); therefore, progressive distillation can reduce nationwide GHG emissions from petroleum refineries by almost 5 percent.

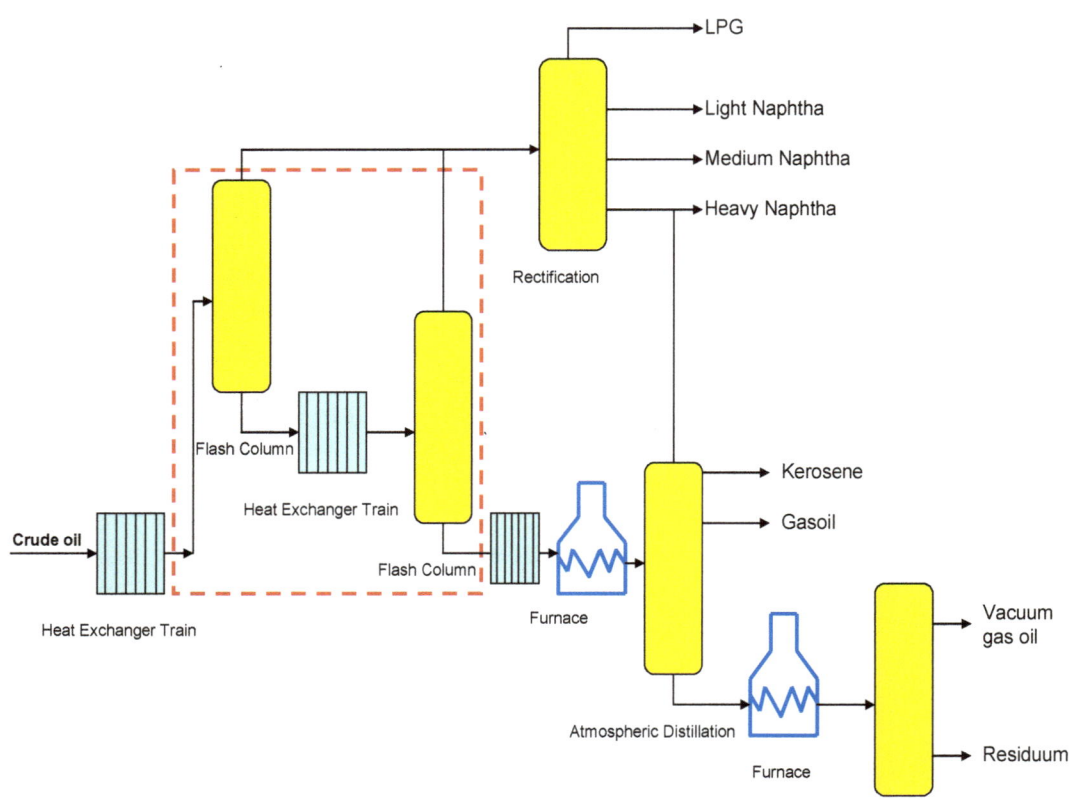

5.10 Storage Tanks

5.10.1 Vapor Recovery or Control for Unstabilized Crude Oil Tanks

Crude oil often contains methane and other light hydrocarbons that are dissolved in the crude oil because the crude oil is "stored" within the wells under pressure. When the crude oil is pumped from the wells and subsequently stored at atmospheric pressures, CH_4 and other light hydrocarbons are released from the crude oil and emitted from the atmospheric storage tanks. Most refineries receive crude oil that has been stored for several days to several weeks at atmospheric pressures prior to receipt at the refinery. These stabilized crude oils have limited GHG emissions. If a refinery receives crude oil straight from a production well via pipeline without being stored for several days at atmospheric pressures, the crude oil may contain significant quantities of methane and light VOC. When this "unstabilized" crude oil is first stored at the refinery at atmospheric conditions, the methane and gaseous VOC will evolve from the crude oil. Common tank controls, such as floating roofs, are ineffective at reducing these emissions. If a refinery receives unstabilized crude oil, a fixed roof tank vented to a gas recovery system of control device could be considered to reduce the GHG (particularly CH_4) emissions from these tanks.

5.10.2 Heated Storage Tank Insulation

Some storage tanks are heated to control viscosity of the stored product. A study at a refinery found that insulating an 80,000 bbl storage tank that is heated to 225 °F could save $148,000 in energy costs (Worrell and Galitsky, 2005).

6.0   References

1. ARCADIS. 2008. Air Pollution Control and Efficiency Improvement Measures for U.S. Refineries. Prepared for U.S. Environmental Protection Agency, Research Triangle Park, NC. Contract No. EP-C-04-023. August 8.

2. Barker, D.J., S.A. Turner, P.A. Napier-Moore, M. Clark, and J.E. Davison, 2009. "CO2 Capture in the Cement Industry," Energy Procedia, Vol. 1, pp. 87-94. http://www.sciencedirect.com/science?_ob=MImg&_imagekey=B984K-4W0SFYG-F-1&_cdi=59073&_user=10&_orig=browse&_coverDate=02%2F28%2F2009&_sk=999989-998&view=c&wchp=dGLbVlz-zSkWA&md5=12853bece66323782f9a46335b4b213c&ie=/sdarticle.pdf

3. Bosoaga, Adina, Ondrej Masek, and John E. Oakey. 2009. CO_2 Capture Technologies for Cement Industry, Energy Procedia, Vol. 1, pp. 133-140. http://www.sciencedirect.com/science?_ob=MImg&_imagekey=B984K-4W0SFYG-N-1&_cdi=59073&_user=10&_orig=browse&_coverDate=02%2F28%2F2009&_sk=999989998&view=c&wchp=dGLzVlz-zSkWA&md5=b9ba07fff56e3ad43069658cb689db9e&ie=/sdarticle.pdf

4. Coburn J. 2007. Greenhouse Gas Industry Profile for the Petroleum Refining Industry. Prepared for U.S. Environmental Protection Agency, Washington, DC. Contract No. GS-10F-0283K. June 11.

5. Ecofys. 2009. Methodology for the free allocation of emission allowances in the EU ETS post 2012. Sector report for the refinery industry. November.

6. ECRA (European Cement Research Academy). 2009. Development of State of the Art – Techniques in Cement Manufacturing: Trying to Look Ahead, June 4, 2009, Düsseldorf, Germany.), Cement Sustainability Initiative. http://www.wbcsdcement.org/pdf/technology/Technology%20papers.pdf

7. EIA (Energy Information Administration). 2006. Refinery Capacity Report 2006. Prepared by the Energy Information Administration, Washington, DC. June 15.

8. EIA (Energy Information Administration). 2009. Refinery Capacity Report 2008. Prepared by the Energy Information Administration, Washington, DC. June 25. See on-line HTML version of Table 12a: http://tonto.eia.doe.gov/dnav/pet/pet_pnp_capfuel_dcu_nus_a.htm.

9. Envirocomb Limited. 2006. Zero flaring by flare gas recovery. Presented at the Gas Processors Association 1st Specialized Technical Session: Zero Flaring Seminar. Al-

Khobar, Saudi Arabia, November 29. Available at: http://www.gpa-gcc-chapter.org/PDF/Zero%20Flaring%20By%20Flare%20Gas%20Recovery.pdf

10. Gary, J. H. and G.E. Handwerk. 1994. Petroleum Refining Technology and Economics. 3rd Edition. Marcel Dekker, Inc. New York, NY.

11. ICF International. 2010. CHP Installation Database, Maintained for U.S. DOE and Oak Ridge National Laboratory.

12. IPCC (International Panel on Climate Change). 2006. *2006 IPCC Guidelines for National Greenhouse Gas Inventories*. Prepared by the National Greenhouse Gas Inventories Programme. Edited by H.S. Eggleston, L. Buendia, K. Miwa, T. Ngara, and K. Tanabe. IGES: Japan.John Zinc Company. 2006. Flare gas recovery (FGR) to reduce plant flaring. Presented at the Gas Processors Association 1st Specialized Technical Session: Zero Flaring Seminar. Al-Khobar, Saudi Arabia, November 29. Available at: http://www.gpa-gcc-chapter.org/PDF/Flare%20Gas%20Recovery%20White%20Paper.pdf.

13. Lucas, Bob. 2008. Memorandum to Petroleum Refinery New Source Performance Standards (NSPS) Docket No. EPA-HQ-OAR-2007-0011 from Bob Lucas, USEPA regarding Documentation of Flare Recovery Impact Estimates. April 26.

14. Peterson, J, N. Tuttle, H. Cooper, and C. Baukal. 2007. Minimize facility flaring. Hydrocarbon Processing. June. Pp. 111-115. Available at: http://www.johnzink.com/products/flares/pdfs/flare_hydro_proc_june_2007.pdf.

15. U.S. DOE (Department of Energy). 2002. Steam System Opportunity Assessment for the Pulp and Paper, Chemical Manufacturing, and Petroleum Refining Industries. U.S. Department of Energy, DOE/GO-102002-1640. October 2002.

16. U.S. DOE (Department of Energy). 2005. Petroleum Best Practices Plant-wide Assessment Case Study. Valero: Houston refinery uses plant-wide assessment to develop an energy optimization and management system. DOE/GO-102005-2121. Available at: http://www1.eere.energy.gov/industry/bestpractices/pdfs/valero.pdf

17. U.S. DOE (Department of Energy). 2007. Energy and Environmental Profile of the U.S. Petroleum Refining Industry. Prepared by Energetics, Inc., Columbia, MD. November 2007.

18. U.S. EPA (Environmental Protection Agency). 1995. EPA Office of Compliance Sector Notebook Project: Profile of the Petroleum Refining Industry. EPA/310-R-95-013. Washington, DC. September 1995.

19. U.S. EPA (Environmental Protection Agency). 1998. Petroleum Refineries-Background Information for Proposed Standards, Catalytic Cracking (Fluid and Other) Units, Catalytic Reforming Units, and Sulfur Recovery Units. EPA-453/R-98-003. Washington, DC: Government Printing Office.

20. U.S. EPA (Environmental Protection Agency). 1999. U.S. Methane Emissions 1990–2020: Inventories, Projections, and Opportunities for Reductions. EPA 430-R-99-013. Section 3. Natural Gas Systems and Appendix III. Washington, DC. September 1999.

21. U.S. EPA (Environmental Protection Agency). 2008. Technical Support Document for the Petroleum Refining Sector: Proposed Rule for Mandatory Reporting of Greenhouse Gases. Docket Item EPA-HQ-OAR-2008-0508-0025. Office of Air and Radiation. Office of Atmospheric Programs, Washington, DC. September 8.

22. Worrell, Ernst and Christina Galitsky. 2005. Energy Efficiency Improvement and Cost Saving Opportunities for Petroleum Refineries (Report No. LBNL-56183). Ernest Orlando Lawrence Berkeley National Laboratory, Berkeley, CA. February 2005. http://www.energystar.gov/ia/business/industry/ES_Petroleum_Energy_Guide.pdf

23. Worrell, Ernst and Christina Galitsky. 2008. Energy Efficiency Improvement and Cost Saving Opportunities for Cement Making (Report No. LBNL-54036-Revision). Ernest Orlando Lawrence Berkeley National Laboratory, Berkeley, CA. March 2008. http://www.energystar.gov/ia/business/industry/LBNL-54036.pdf

EPA Contact

Brenda Shine
U.S. EPA
OAQPS/SPPD/CCG
Mail Code E143-01
Research Triangle Park, NC 27711
Phone: 919-541-3608
shine.brenda@epa.gov

www.ingramcontent.com/pod-product-compliance
Lightning Source LLC
Chambersburg PA
CBHW050354180526
45159CB00005B/2012

9 7 8 1 5 0 0 9 5 4 0 9 3